Elephant, Deer, and Other Herds

Cynthia O'Brien

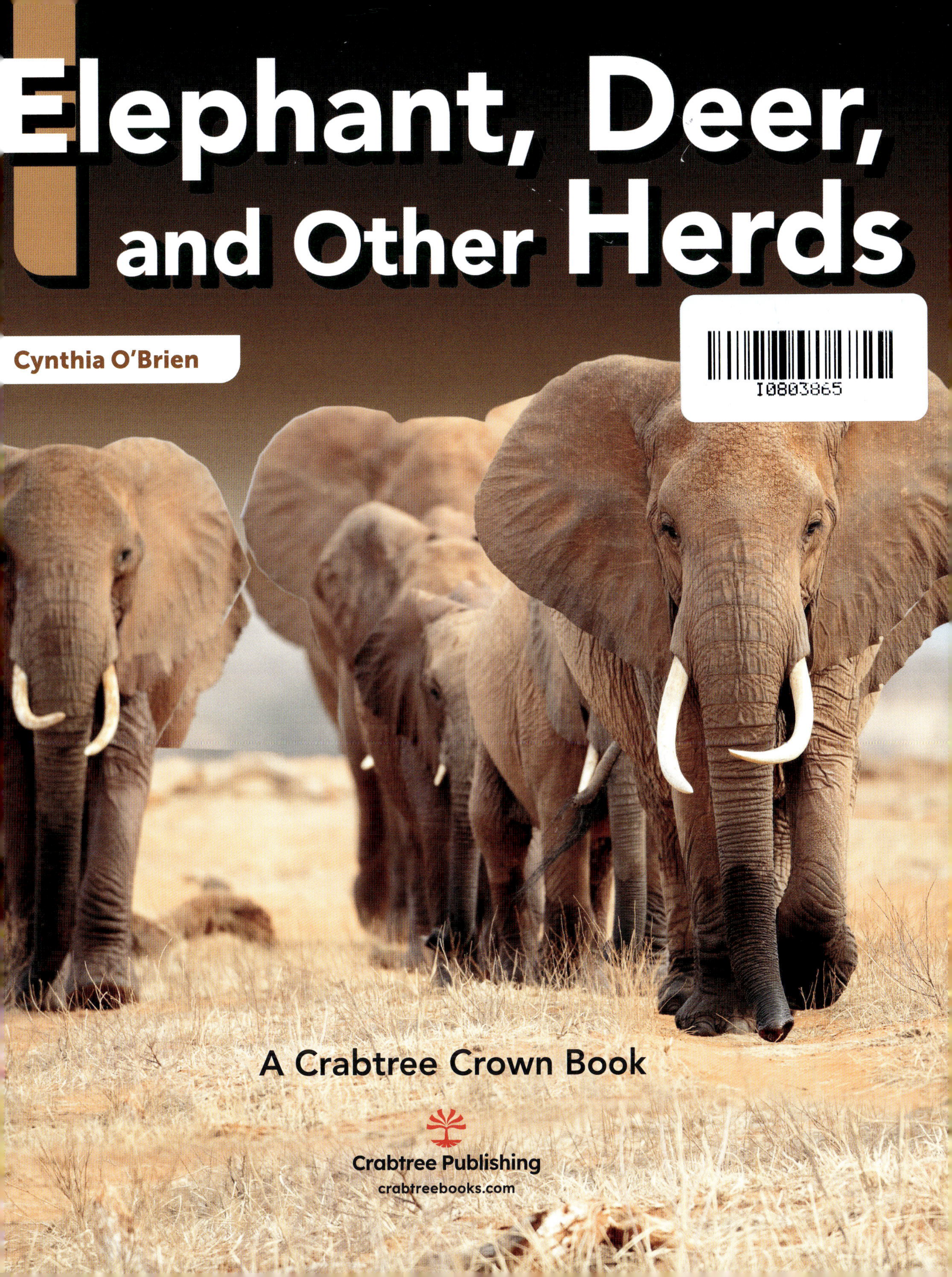

A Crabtree Crown Book

Crabtree Publishing
crabtreebooks.com

Author: Cynthia O'Brien

Series research and development:
Reagan Miller, Janine Deschenes

Editor: Janine Deschenes

Proofreader: Kathy Middleton

Design and photo research:
Katherine Berti

Print and production coordinator:
Katherine Berti

Image credits:
Alamy Stock Photo
robertharding: p. 29 (bottom right)
Flickr
GPA Photo Archive, Tom Koerner, USFWS: p. 21 (bottom)
All other images from Shutterstock

Crabtree Publishing

crabtreebooks.com 800-387-7650

Hardcover 978-1-0398-0652-8
Paperback 978-1-0398-0678-8
Ebook (pdf) 978-1-0398-0703-7
Epub 978-1-0398-0730-3

Published in Canada
Crabtree Publishing
616 Welland Avenue
St. Catharines, Ontario
L2M 5V6

Published in the United States
Crabtree Publishing
347 Fifth Avenue
Suite 1402-145
New York, New York, 10016

Library and Archives Canada Cataloguing in Publication
Available at Library and Archives Canada

Library of Congress Cataloging-in-Publication Data
Available at the Library of Congress

Printed in the U.S.A./012023/CG20220815

Contents

What Is a Herd?

Think of a farm. What kinds of animals are found there? Did you imagine a large group of cows, sheep, or horses together in a pasture? These animals live in groups called herds.

*Herds may be **domestic** or wild. Some farm animals live in domestic herds. People feed them and look after them.*

Many herds are made up of hoofed animals, such as deer. But other animals, such as elephants and walruses, also form herds.

Herd animals are often hunted as ***prey***.

Living in a group helps animals survive. Animal groups can be large or small. A small group might include family members only. Some animals may live in a group all year. Other animals form groups to **socialize**, hunt, or **migrate**.

The name for an animal group depends on the animal. Wolves live in groups called packs. Gorillas and other apes live in troops. Many fish swim in schools. Deer, bison, cows, sheep, and other animals live in herds.

Why Do Animals Live in Herds?

Young animals learn from adults in a herd. They discover how to watch for **predators**. They find out the best spots for food.

Herds may be all female, all male, or a mix. There are young male groups, called bachelor herds. Nursery herds are made up of a few adult females and their young.

Some herds have a leader. In an elephant herd, the oldest and strongest female is the leader.

When young animals are old enough, they sometimes leave and form a new herd.

caribou

All animals in a herd rely on each other for protection. As a group, herd animals can provide **camouflage** for each other. Predators may find it difficult to see individual animals. Living in a herd also means that there are many animals to keep watch for predators.

If a herd animal spots danger, it warns the rest of the herd. For example, white-tailed deer snort loudly to give alarm signals.

Elephants

At a Glance

Number of species: Two—African elephant and Asian elephant
Herd size: 10 or more
Location: Africa and Asia

African elephants are larger than their Asian cousins. They also have much larger ears, which help keep them cool.

Elephants are the largest land **mammals** in the world. The oldest female leads the elephant herd. Other females and young elephants make up the rest of the herd. The leader teaches her herd. She shows them where to find food and water. She helps keep them safe.

Female elephants usually stay with their mother's herd their whole lives. Young males leave the herd when they are between 10 and 14 years old. They form or join herds with other males. These bachelor herds change as the males go off on their own or meet up with other male herds.

Did You Know?

Asian elephants use their trunks to comfort each other. If an elephant is upset, another elephant will put its trunk in the upset elephant's mouth. This is like an elephant hug. Elephants also make chirping sounds to calm each other.

Elephant herds migrate each year to areas with more food and water. They always follow the same routes.

Elephants also use their trunks to make warning sounds, say hello to other elephants, pick up things, and suck up water.

African Elephants

African elephants live in all kinds of **habitats**. They roam in grasslands, deserts, and forests. Herds get together in clearings and near water to eat and socialize. All elephants are **herbivores**. These huge animals spend a lot of time looking for and eating food.

Elephant herds sometimes gather to form a large group of females and their young. This group may include hundreds of members. Adult female elephants work together to take care of the baby elephants, or calves. Many will respond if a calf cries out. They always keep a careful watch for predators.

Elephant herds are mostly made up of females and their calves. Males, or bull elephants, sometimes join the herds for a short time.

Lions and hyenas sometimes attack young elephants. But human hunters are the greatest threat to African elephants. They illegally hunt elephants for their ***ivory*** *tusks.*

Sticking Together

Living in a herd means there are always other elephants to play with. A young elephant lifts up its trunk to invite other elephants to play. They like to run and **trumpet** for fun. Males also play fight, but an older sister will keep watch to make sure no one gets hurt.

Elephants may trumpet when they are excited, surprised, angry, or playful.

Asian Elephants

Asian elephants are **endangered** in the wild. About one third of all Asian elephants live in **captivity**. Wild herds live in forested areas. They migrate to find food and water. They gather with other herds in places where there is a lot of food.

Elephants communicate with each other using many sounds. Growls, trumpets, barks, and rumblings all have different meanings. The leader rumbles and flaps her ears to signal that it is time to move to a different place. Low rumbles and stamping feet mean danger is near. Communication helps keep the herd together and safe.

Elephants can pick up sounds and movements in the ground as far as 20 miles (32 km) away.

Sticking Together

Elephants do not leave members of the herd behind. If an elephant is hurt, the other elephants try to help. If an older or injured elephant cannot move very quickly, the whole herd slows down.

Asian elephants are found in India and countries in Southeast Asia, such as Thailand.

Deer

At a Glance

Number of species: At least 43
Herd size: Two to many thousands
Location: Worldwide, except Antarctica

The deer family includes white-tailed deer, caribou, elk, and many others. Most deer species live in herds. Sometimes females have their own herd and males live in another herd. If a herd is made up of males and females, the strongest male becomes the leader. Herds of deer can have just a few members or thousands.

Moose are the largest deer, but they usually live alone. They come together only when it is time to ***mate****.*

All deer have many predators, including bears and wolves. Deer that live in herds are better protected. Adult deer are very watchful. They have excellent hearing, eyesight, and sense of smell. If a deer senses danger, it warns the rest of the herd.

The size of the herd depends on the type of deer and the season. Some deer, such as elk (below), migrate together in large herds to find food and warmer weather.

Caribou

At a Glance

Number of subspecies: Seven, including barren-ground
Herd size: 10 to many thousands
Location: Northern areas of North America, Europe, Asia, and Greenland

Caribou is a species of deer. In North America, barren-ground caribou live in very large herds. They have thousands of animals. These large herds sometimes mix with others in the summer. Woodland caribou live in smaller herds and do not migrate.

Living in a herd helps keep caribou safe. Caribou keep constant watch for predators such as wolves and bears. They have special scent **glands** on their ankles. The scent glands create a smell if a caribou spots danger. This smell alerts the others in the herd.

Peary caribou are the smallest ***subspecies*** *of caribou. They are mostly white in the winter. This gives them better camouflage in their Arctic habitat.*

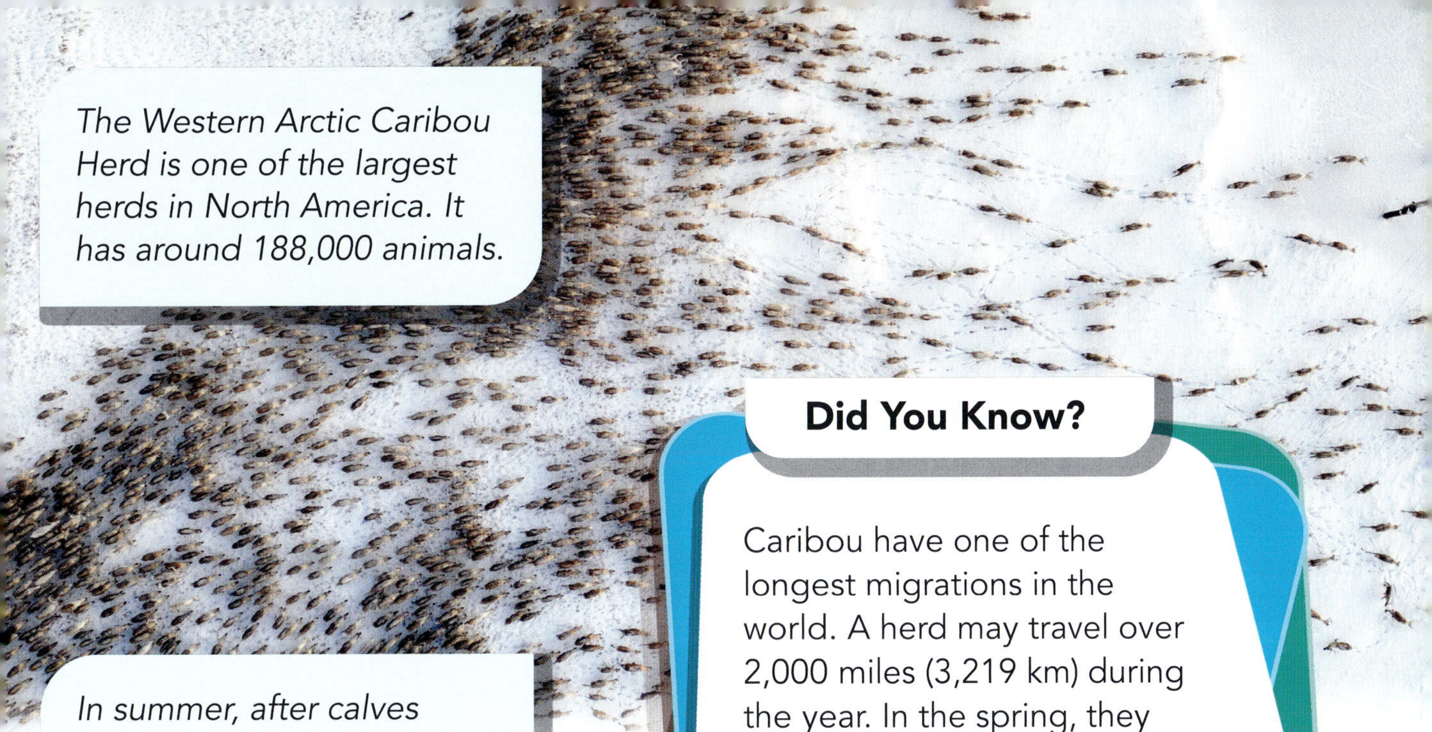

The Western Arctic Caribou Herd is one of the largest herds in North America. It has around 188,000 animals.

In summer, after calves are born, caribou gather in huge groups. This helps give protection from pests such as mosquitos and flies.

Did You Know?

Caribou have one of the longest migrations in the world. A herd may travel over 2,000 miles (3,219 km) during the year. In the spring, they move north to calving grounds where new caribou are born. In the winter, they move south again.

White-tailed Deer

At a Glance

Number of subspecies: Nearly 40
Herd size: Two to 12
Location: Southern Canada, United States, Central America

The white-tailed deer is the smallest deer in North America. Males and females live separately for much of the year. They come together to socialize and eat in the summer. Then they meet again in the fall for **breeding** season. When they are all together, the largest adult males are the most important in the herd. Next are the adult females and last are the young deer.

The adult female leader teaches young deer skills such as where to find food. She also snorts or stamps her foot to warn the herd of danger.

Fawns need to stay hidden while their mothers search for food. Their spotted coats give camouflage. They also have very little scent.

Did You Know?

White-tailed deer run as fast as 30 miles per hour (48 kph). They can also jump as far as 30 feet (9 m). These skills help them escape from predators.

Females, their female relatives, and their young live in small herds for much of the year. Young male deer leave their mothers' herds when they are around one year old. They join up with adult male herds or other young males.

White-tailed deer raise their tails when they sense danger. The flash of white color helps alert the herd.

Antelopes

Gazelles, impalas, and wildebeest are just a few of many antelope species. Many of them live in herds, from very small families to extremely large groups.

Antelopes live together for protection from predators, to migrate, and to breed. In some herds, animals stand guard to alert the rest of the animals to any danger. Many antelopes have nursery herds to keep the young safe. Adults protect the young, sometimes using their horns.

At a Glance

Number of species: About 135
Herd size: Two to many thousands
Location: Africa and Asia

Antelopes have many predators, including cheetahs, crocodiles, and lions.

Though antelope species look different from each other, all have horns. In some species, such as the impala (below), only males have horns.

Pronghorns

At a Glance

Number of subspecies: Five
Herd size: 10 to hundreds
Location: North America, from southern Canada to Mexico

The pronghorn is unique to North America. Most pronghorns live in small herds in warmer weather. Females form small groups with their young, called fawns. Young males gather in separate herds.

In winter, pronghorns in colder areas travel to find food. Hundreds of male and female pronghorns gather to form a migrating herd.

Did You Know?

Pronghorns have excellent eyesight. This is useful for spotting danger and for seeing signals. If a predator is nearby, a pronghorn will raise the white hairs on its rump. This warns other pronghorns in the herd to run away.

The pronghorn is one of the fastest land animals. It can run nearly 60 miles per hour (97 kph).

Zebras

At a Glance

Number of species: Three—Plains, Mountain, and Grevy's
Herd size: Five to 20
Location: Grasslands and woodlands of southern and eastern Africa

With their black-and-white striped coats, zebras are easy to identify. Most zebras, other than Grevy's, live in family herds. These herds have one adult male, several females, and their young. The adult male, or stallion, leads the herd. The stallion uses powerful kicks to protect the other zebras from predators such as hyenas or lions.

Sometimes hundreds of zebras gather in places where there is plenty of food and water. Thousands of zebras may travel together for long migrations. But the zebra family always stays close together. If a zebra gets lost, the others will look for it.

Zebras look similar to one another, but each animal has its own, unique pattern. A herd of zebras is also called a dazzle. The name describes the movement of the stripes when zebras run in a group.

Sticking Together

Zebras are very social animals. They communicate with each other with snorts and different facial expressions. They **groom** each other and play by chasing or pretending to fight.

A zebra might show its teeth to threaten another animal.

Zebras often migrate with other animals, especially wildebeest. This buddy system helps both groups stay safer from predators.

Giraffes

At a Glance

Number of species: Four—Masai, Southern, Northern, Reticulated
Herd size: Three to 20
Location: Africa, south of the Sahara Desert

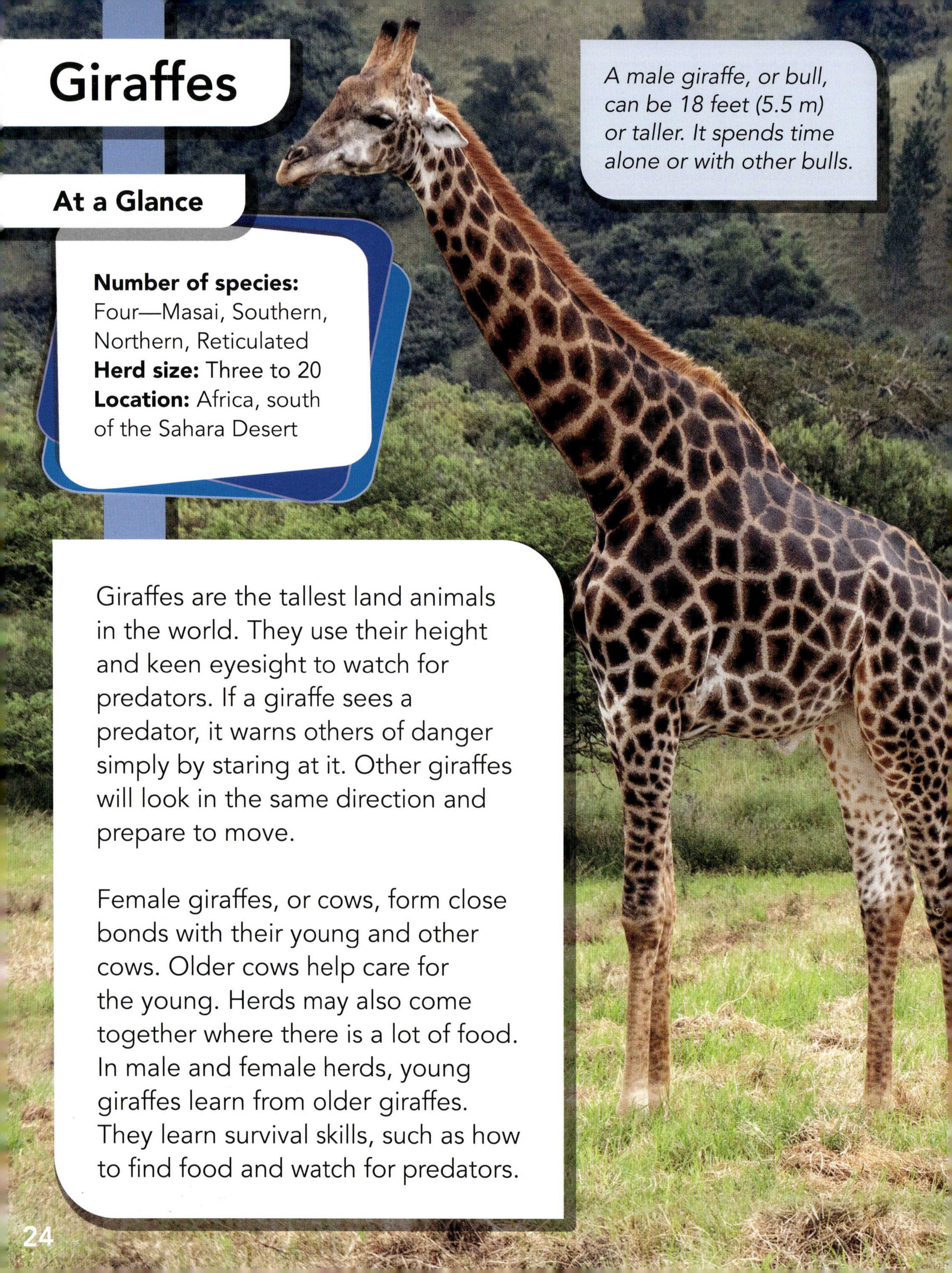

A male giraffe, or bull, can be 18 feet (5.5 m) or taller. It spends time alone or with other bulls.

Giraffes are the tallest land animals in the world. They use their height and keen eyesight to watch for predators. If a giraffe sees a predator, it warns others of danger simply by staring at it. Other giraffes will look in the same direction and prepare to move.

Female giraffes, or cows, form close bonds with their young and other cows. Older cows help care for the young. Herds may also come together where there is a lot of food. In male and female herds, young giraffes learn from older giraffes. They learn survival skills, such as how to find food and watch for predators.

Sticking Together

Young male giraffes form bachelor herds. This gives them time for learning and play. They also battle to find out who is strongest. Two giraffes swing their necks at each other. This is called necking.

When it is time to find food, mother giraffes leave their calves in nursery groups. One cow stays behind to watch the group and protect the young from predators.

When they are necking, giraffes use their heads as clubs. They do not seriously hurt each other.

Bison

At a Glance

Number of species: Two—European and American
Herd size: 20 to 30
Location: Europe, especially Poland and Belarus, and protected or private areas in the U.S. and Canada

Bison are very large herbivores. American bison live in grasslands. European bison prefer woodlands. Both species live in herds.

Most of the year, females and adult males live apart. One or more females, or cows, form herds with young bison. The oldest cow leads the herd, but all the adults help take care of the young. They protect them and each other from wolves and bears. Meanwhile, adult males, or bulls, live alone or in a herd with other males. The oldest and largest bull is the leader.

The bison is the largest land animal in North America. A male can weigh more than 2,000 pounds (907 kg).

Did You Know?

Many millions of American bison used to roam **plains** in North America. In the 1800s, people hunted the bison in great numbers. By 1889, there were just 1,000 bison left. Today, about 400,000 American bison live on protected lands and ranches.

In the summer, breeding season can bring 1,000 bison together in one area. Males often butt heads and growl at each other to impress females.

Bison migrate between summer and winter to find areas with more food.

Walruses

At a Glance

Number of species: Two—Atlantic and Pacific
Herd size: Over 100 to many thousands
Location: Coastal Arctic areas in Canada, Russia, and Alaska

Walruses are related to seals, but have large teeth called tusks. They use their tusks to pull their heavy bodies out of the sea and onto rocks, ice, or beaches. This is called hauling out. Walruses haul out as a group and rest close together.

Males and females usually have separate, large herds. They come together when it is time to mate. Thousands of walruses pile together on the rocks. They bark and grunt at one another to communicate. Adult males use their tusks to fight each other. They do this to prove their importance and to compete for female walruses.

Unlike most herd animals, walruses spend their lives near icy seas.

Sticking Together

Young walruses, or calves, are prey for orcas and polar bears. Many female walruses huddle with their calves to protect them. Walruses also protect injured animals in their herd. They keep them away from hunters and try to get them to safety.

Walruses eat seafood such as shrimp and clams.

Walrus calves often stay safe by riding on their mothers' backs.

Glossary

breeding Having babies

camouflage An animal's physical characteristics that allow it to blend into its surroundings

captivity Of animals, living under human care or control

domestic Describes animals kept by humans for work or as pets

endangered At risk of extinction

glands Organs that make certain substances, such as sweat

groom For animals, an often social behavior in which one animal cleans another's body

habitats Natural homes or environments of animals

herbivores Animals that eat plants

ivory A hard, white substance that makes up animal tusks

mammals Warm-blooded animals with backbones. Mammal babies drink milk from their mothers' bodies

mate To join together to make babies

migrate Of animals, move from one place to another, often according to the season

plains Large, flat areas of land with few trees

predators Animals that hunt other animals for food

prey Animals that are hunted by other animals for food

socialize Living and interacting with others in a group

subspecies Two or more distinct categories of one species

trumpet A loud call, made especially by an elephant

Learning More

On the Web

www.natgeokids.com/uk/discover/animals/general-animals/zebra-facts
Find out interesting facts about zebras and zebra herds, and look up other animal groups.

https://kids.britannica.com/kids/article/elephant/353093
Discover information about elephants and their herds, and watch informational videos about these massive mammals.

www.adfg.alaska.gov/index.cfm?adfg=caribou.main
Explore all kinds of information about caribou, from the sounds they make to threats they face.

Books

Kalman, Bobbie. *Animals that Live in Social Groups*. Crabtree Publishing, 2016.

Newland, Sonya. *Epic Migrations by Land*. Crabtree Publishing, 2020.

Santos, Tracie. *Animal Societies*. Discovery Library, 2021.

Index

About the Author

Cynthia O'Brien has written many books for children, including books about animals and the natural world. Researching this book, she learned a lot about how different animals live together and help each other.